~A BINGO BOOK~

Crime-Scene Investigation Bingo Book

COMPLETE BINGO GAME IN A BOOK

Written By Rebecca Stark

ISBN 978-0-87386-428-2

© 2016 Barbara M. Peller

Educational Books 'n' Bingo

Printed in the U.S.A.

CRIME-SCENE INVESTIGATION BINGO
Directions

INCLUDED:

List of Terms

Templates for Additional Terms and Clues

2 Clues per Term

30 Unique Bingo Cards

Markers

1. **Either cut apart the book or make copies of ALL the sheets. You might want to make an extra copy of the clue sheets to use for introduction and review. Keep the sheets in an envelope for easy reuse.**

2. Cut apart the call cards with terms and clues.

3. Pass out one bingo card per student. There are enough for a class of 30.

4. Pass out markers. You may cut apart the markers included in this book or use any other small items of your choice.

5. Decide whether or not you will require the entire card to be filled. Requiring the entire card to be filled provides a better review. However, if you have a short time to fill, you may prefer to have them do the just the border or some other format. Tell the class before you begin what is required.

6. There are 50 topics. Read the list before you begin. If there are any topics that have not been covered in class, you may want to read to the students the topic and clues before you begin.

7. There is a blank space in the middle of each card. You can instruct the students to use it as a free space or you can write in answers to cover topics not included. Of course, in this case you would create your own clues. (Templates provided.)

8. Shuffle the cards and place them in a pile. Two or three clues are provided for each topic. If you plan to play the game with the same group more than once, you might want to choose a different clue for each game. If not, you may choose to use more than one clue.

9. Be sure to keep the cards you have used for the present game in a separate pile. When a student calls, "Bingo," he or she will have to verify that the correct answers are on his or her card AND that the markers were placed in response to the proper questions. Pull out the cards that are on the student's card keeping them in the order they were used in the game. Read each clue as it was given and ask the student to identify the correct answer from his or her card.

10. If the student has the correct answers on the card AND has shown that they were marked in response to the *correct questions,* then that student is the winner and the game is over. If the student does not have the correct answers on the card OR he or she marked the answers in response to *the wrong questions,* then the game continues until there is a proper winner.

11. If you want to play again, reshuffle the cards and begin again.

Have fun!

TERMS

AFIS

antemortem

autopsy

ballistics

barrier tape

blood splatter

caliber

cast

cast-off blood

cause of death

chain of custody

chromosome

class evidence

CODIS

cold case(s)

contamination

coroner(s)

corpus delicti

criminalistics

DNA

Edmond Locard

entomology (-gist)

evidence

fingerprint(s)

FISH

forensic

forgery

gas chromatograph

gunshot residue

handwriting

height

impression evidence

indented writing

latent

Latin

ligature

luminol

manner(s) of death

medical examiner (ME)

microscope(s)

modus operandi (MO)

mortis

odontology

pH scale

postmortem

rigor mortis

suspect

tool marks

trace evidence

trajectory

Additional Terms

Choose as many other terms as you would like and write them in the squares.
Repeat each term as many times as desired.
Cut out the squares and randomly distribute them to the class.
Instruct the students to place the square on the center space of their card.

Crime Scene Investigation Bingo

Clues for Additional Terms

Write two clues for each of your additional terms.

_____	_____
1. 2.	1. 2.
_____ 1. 2.	_____ 1. 2.
_____ 1. 2.	_____ 1. 2.

CRIME SCENE DO NOT CROSS
CRIME SCENE DO NOT CROSS
CRIME SCENE DO NOT CROSS
CRIME SCENE DO NOT CROSS
CRIME SCENE DO NOT CROSS

CRIME SCENE DO NOT CROSS
CRIME SCENE DO NOT CROSS
CRIME SCENE DO NOT CROSS
CRIME SCENE DO NOT CROSS
CRIME SCENE DO NOT CROSS

CRIME SCENE DO NOT CROSS
CRIME SCENE DO NOT CROSS
CRIME SCENE DO NOT CROSS
CRIME SCENE DO NOT CROSS
CRIME SCENE DO NOT CROSS

CRIME SCENE DO NOT CROSS
CRIME SCENE DO NOT CROSS
CRIME SCENE DO NOT CROSS
CRIME SCENE DO NOT CROSS
CRIME SCENE DO NOT CROSS

CRIME SCENE DO NOT CROSS
CRIME SCENE DO NOT CROSS
CRIME SCENE DO NOT CROSS
CRIME SCENE DO NOT CROSS
CRIME SCENE DO NOT CROSS

CRIME SCENE DO NOT CROSS
CRIME SCENE DO NOT CROSS
CRIME SCENE DO NOT CROSS
CRIME SCENE DO NOT CROSS
CRIME SCENE DO NOT CROSS

CRIME SCENE DO NOT CROSS
CRIME SCENE DO NOT CROSS
CRIME SCENE DO NOT CROSS
CRIME SCENE DO NOT CROSS
CRIME SCENE DO NOT CROSS

AFIS 1. Acronym for Automated Fingerprint Identification System. 2. This system scans fingerprints electronically and compares them with prints in a database.	**antemortem** 1. Usually used as an adjective, it means "before death." 2. Complete this analogy: before : after :: ___ : postmortem.
autopsy 1. The internal and external examination of a body after death. 2. An ___ is performed to confirm or determine the cause of death and establish other pre-death conditions.	**ballistics** 1. ___ is the study of the motion of bullets and their examination for distinctive characteristics after being fired. 2. Examiners can use ___ evidence to match bullets or bullet fragments to specific weapons.
barrier tape 1. ____ is essential for establishing a line of security around a crime scene. 2. ____ and numbered evidence markers are important crime-scene supplies.	**blood splatter** 1. The pattern of blood that has struck a surface is called ___. 2. This pattern can provide vital information about the source of the blood.
caliber 1. The diameter of the bore of a rifled firearm is its ___. 2. The ___ of a firearm is usually expressed in hundredths of an inch or in millimeters.	**cast** 1. An impression, such as a shoe print, can be lifted from the ground by making a ___. 2. A crime-scene investigator can compare a ___ of a shoe print with the bottom of a suspect's shoe.
cast-off blood 1. Droplets of blood flung from the object so as to make a trail of blood where it lands is ___. 2. A moving source of blood, such as a bleeding victim or a bloody weapon, can give rise to ___.	**cause of death** 1. The action that resulted in death is called the ___. 2. A blow to the head or a bullet wound to the head can be the ___.

chain of custody	chromosome
1. A list that records every official person who handles a piece of evidence. 2. Persons in the ___ put their initials and the date on the evidence container.	1. It is a single piece of coiled DNA containing many genes. 2. A gene is a sequence of DNA that occupies a specific location on a ___ and determines a particular trait.
class evidence	**CODIS**
1. Evidence that can be linked to a group of people but is not specific enough to be linked to an individual. 2. It is evidence that is specific enough to identify overall characteristics but too general for a unique identification.	1. This acronym stands for Combined DNA Index System. 2. This system is used to share DNA profiles kept in the FBI's index system with law enforcement bodies.
cold case(s)	**contamination**
1. An old, unsolved criminal case is a ___. 2. Many ___ are now being solved with the advent of DNA testing.	1. ___ is the act of ruining evidence by accidentally depositing outside trace evidence on items from a crime scene or a suspect. 2. Crime scene ___ usually results through the actions of the personnel at the crime scene.
coroner(s)	**corpus delicti**
1. The ___ is a public official responsible for investigating when someone dies unexpectedly or suspiciously. The ___ may or may not be a doctor. 2. Only some ___ are trained to be forensic pathologists; ___ are elected or appointed to their positions.	1. ___ evidence is evidence that a crime has been committed. It is a Latin term meaning "body of crime." 2. In a murder case, the ___ is the dead body of the victim.
criminalistics	**DNA**
1. ___ is the forensic science of analyzing and interpreting evidence. 2. Criminology explores the nature of crime and prevention; ___ involves the collection and analysis of evidence.	1. This acronym stands for deoxyribonucleic acid. It forms chains of genetic material organized into chromosomes. 2. ___ forms chains of genetic material organized into chromosomes.

Edmond Locard 1. He is known for the principle named for him; it states that whenever 2 subjects come into contact with one another, materials are exchanged between them. 2. He is best known for his Exchange Principle.	**entomology (-gist)** 1. Forensic ___ is the study of insects and arthropods in relation to a criminal investigation. 2. "Bugs" on a corpse can tell a forensic ___ how long the body has been dead.
evidence 1. Crime-scene ___ includes anything used, left, removed, altered, or contaminated during the commission of the crime. 2. Personal ___ includes eyewitness accounts. Physical ___ includes fingerprints, fibers and hairs.	**fingerprints** 1. Loops, ridges and whorls are all basic ___ patterns. 2. To make ___ visible, scientists dust surfaces with powder.
FISH 1. This acronym stands for Forensic Information System for Handwriting. 2. Meaning Forensic Information System for Handwriting, ___ is a data-base for handwriting samples.	**forensic** 1. ___ science, or ___s is the application of science to the criminal and civil laws that are enforced by police agencies in a criminal justice system. 2. The word ___ can be used as a synonym for *legal.*
forgery 1. A ___ is something false used to deceive. 2. If you sign a document with someone else's name without permission, it is a ___.	**gas chromatograph** 1. A ___, or GC, is a forensic tool used to identify the chemical makeup of substances used in the commission of crimes. 2. With a ___, the substance being identified is burned at high temperatures. The temperature at which it becomes gas determines its makeup.
gunshot residue 1. Unburned primer powder sprayed on to the hands of someone firing a gun is known as ___. 2. ___ is evidence that the person fired a gun.	**handwriting** 1. Two basic ___ characteristics are connecting strokes and letter formation. 2. When comparing ___ samples, an expert would notice if letters end with a flourish.

Crime Scene Investigation Bingo

height 1. A person's ___ can be determined by the length of the tibia, fibula or humerus. 2. If a man's tibia is 43 centimeters, his ___ is probably about 6 feet.	**impression evidence** 1. ___, such as tire tracks, can be photographed, lifted with tape, or cast with plaster. 2. Shoe prints and tire tracks are types of ___.
indented writing 1. It is the impression from a writing instrument captured on sheets of paper below the one that contains the original writing. 2. This is a valuable investigation procedure when medical records are suspected of containing alterations.	**latent** 1. ___ means "present, but not visible." 2. A ___ fingerprint is not usually visible to the human eye.
Latin 1. Many terms related to crime-scene investigation come from this ancient language. 2. The word "forensic" comes from the ___ word *forensis,* meaning "of the forum."	**ligature** 1. It is a cordlike object used for strangulation. 2. If ___ marks are present, the cause of death is likely strangulation.
luminol 1. This chemi-luminescent compound reacts to red blood cells and gives off a blue-greenish light. 2. Because ___ reacts with substances other than blood, it cannot be admitted for evidence in court.	**manner(s) of death** 1. This is the legal classification of how someone died as determined by the coroner or medical examiner. 2. Suicide, natural, accidental, or homicide are the 4 possible ___.
medical examiner (ME) 1. A ___ or a coroner determines the cause and manner of death. 2. The ___ determines the cause and manner of the death by performing an autopsy. Unlike a coroner, the ___ is always a trained medical practitioner.	**microscope(s)** 1. It produces a high-magnification image of an object that is otherwise difficult or impossible to see with the unaided eye. 2. A comparison ___ has two compound light ___ with an optical bridge, so two samples can be viewed and compared in a single eyepiece.

Crime Scene Investigation Bingo

modus operandi **(MO)** 1. The usual method of operation used by a perpetrator. 2. It refers to the actions used by the perpetrator to execute the crime, prevent its detection, and/or facilitate escape.	***mortis*** 1. *Algor* ___ is the postmortem cooling of a dead body. 2. *Livor* ___ is a coloration of the skin of the lower parts of a corpse caused by the settling of the red blood cells.
odontology 1. Forensic___ is the application of dentistry to the investigation of crime. 2. Forensic___ has its main applications in identification of corpses and human remains and in bite analysis.	**pH scale** 1. The ___ measures how acidic or basic a substance is. 2. The ___ ranges from 0 to 14 with 7 being neutral.
postmortem 1. It means "after death." It is often used as a synonym for *autopsy.* 2. Complete this analogy: before : antemortem :: after : ___.	***rigor mortis*** 1. ___ is the stiffening of the body that occurs about 30 minutes after death. 2. ___ begins to set in about 30 minutes after death and continues for up to 18 hours.
suspect 1. A ___ is an individual who might possibly have committed the crime under investigation. 2. A ___ is a person of interest who *may have* committed the crime; a perpetrator is the one who actually *did* commit it.	**tool marks** 1. Contact between a tool and another object can create ___ on the object. 2. "Jimmy marks" is a slang term for ___.
trace evidence 1. ___ is material deposited at a crime scene that can only be detected through a deliberate processing procedure. 2. Examples of ___ are hairs and fibers that are accidentally deposited at a crime scene.	**trajectory** 1. The path of a projectile is its ___. 2. A ___ is the path that a moving object follows through space.

Crime Scene Investigation Bingo

Crime Scene Investigation Bingo

evidence	Edmond Locard	fingerprints	trajectory	suspect
cast	AFIS	trace evidence	handwriting	entomology (-gist)
mortis	luminol		FISH	latent
rigor mortis	ballistics	criminalistics	pH scale	forensic
forgery	cast-off blood	chromosome	*corpus delicti*	DNA

Crime Scene Investigation Bingo

rigor mortis	odontology	gas chromatograph	ligature	forgery
forensic	handwriting	blood splatter	ballistics	modus operandi (MO)
medical examiner (ME)	cast-off blood		class evidence	criminalistics
coroner(s)	microscope(s)	luminol	impression evidence	entomology (-gist)
DNA	trace evidence	chromosome	cast	corpus delicti

Crime Scene Investigation Bingo

rigor mortis	criminalistics	handwriting	pH scale	mortis
cast-off blood	AFIS	autopsy	Edmond Locard	height
ballistics	trace evidence		modus operandi (MO)	antemortem
luminol	medical examiner (ME)	forgery	coroner(s)	gas chromatograph
corpus delicti	cast	chromosome	impression evidence	fingerprints

Crime Scene Investigation Bingo

luminol	*modus operandi* (MO)	forgery	cast	fingerprints
gunshot residue	blood splatter	Edmond Locard	ligature	*mortis*
FISH	coroner(s)		suspect	trajectory
criminalistics	contamination	trace evidence	chromosome	autopsy
cold case(s)	DNA	Latin	*corpus delicti*	latent

Crime Scene Investigation Bingo

DNA	suspect	ballistics	blood splatter	cast
gunshot residue	criminalistics	autopsy	class evidence	AFIS
odontology	latent		CODIS	chain of custody
entomology (-gist)	*modus operandi* (MO)	evidence	impression evidence	cold case(s)
handwriting	chromosome	manner(s) of death	luminol	FISH

© Barbara M. Peller

Crime Scene Investigation Bingo

antemortem	*modus operandi* (MO)	gas chromatograph	odontology	latent
pH scale	ballistics	cold case(s)	Edmond Locard	*mortis*
ligature	autopsy		blood splatter	class evidence
chromosome	forgery	impression evidence	Latin	FISH
forensic	criminalistics	evidence	manner(s) of death	fingerprints

Crime Scene Investigation Bingo: Card No. 6

Crime Scene Investigation Bingo

evidence	*modus operandi* (MO)	chain of custody	CODIS	handwriting
forensic	fingerprints	cast-off blood	AFIS	gunshot residue
gas chromatograph	trajectory		class evidence	cause of death
luminol	coroner(s)	*mortis*	*rigor mortis*	medical examiner (ME)
chromosome	cast	impression evidence	Latin	antemortem

Crime Scene Investigation Bingo

FISH	*modus operandi* (MO)	barrier tape	pH scale	cause of death
gunshot residue	odontology	ligature	latent	blood splatter
mortis	indented writing		fingerprints	suspect
corpus delicti	luminol	*rigor mortis*	cold case(s)	coroner(s)
trace evidence	chromosome	Latin	ballistics	forensic

Crime Scene Investigation Bingo

class evidence	handwriting	cast-off blood	*mortis*	latent
cold case(s)	odontology	FISH	ballistics	fingerprints
height	evidence		AFIS	barrier tape
cause of death	DNA	forgery	CODIS	chain of custody
coroner(s)	impression evidence	autopsy	*rigor mortis*	suspect

Crime Scene Investigation Bingo: Card No. 9

© Barbara M. Peller

Crime Scene Investigation Bingo

rigor mortis	pH scale	blood splatter	ligature	manner(s) of death
latent	cause of death	Edmond Locard	AFIS	fingerprints
indented writing	*modus operandi* (MO)		trajectory	medical examiner (ME)
forgery	entomology (-gist)	cold case(s)	impression evidence	height
caliber	forensic	gas chromatograph	DNA	FISH

© Barbara M. Peller

Crime Scene Investigation Bingo

antemortem	*modus operandi* (MO)	ballistics	cold case(s)	forensic
barrier tape	height	CODIS	class evidence	Edmond Locard
gunshot residue	odontology		gas chromatograph	cast-off blood
caliber	*mortis*	impression evidence	cast	*rigor mortis*
autopsy	chromosome	evidence	Latin	handwriting

Crime Scene Investigation Bingo

handwriting	suspect	height	pH scale	class evidence
cast-off blood	trace evidence	odontology	Latin	AFIS
evidence	chain of custody		latent	ligature
chromosome	coroner(s)	fingerprints	*rigor mortis*	gunshot residue
modus operandi (MO)	barrier tape	indented writing	autopsy	cause of death

Crime Scene Investigation Bingo

caliber	suspect	antemortem	height	latent
odontology	barrier tape	*modus operandi* (MO)	class evidence	medical examiner (ME)
pH scale	blood splatter		cast-off blood	chain of custody
FISH	impression evidence	cause of death	indented writing	*rigor mortis*
chromosome	entomology (-gist)	Latin	evidence	CODIS

Crime Scene Investigation Bingo

cast	odontology	ballistics	class evidence	caliber
cause of death	evidence	height	AFIS	*modus operandi* (MO)
cold case(s)	trajectory		gas chromatograph	autopsy
entomology (-gist)	impression evidence	indented writing	blood splatter	antemortem
chromosome	ligature	medical examiner (ME)	forensic	FISH

Crime Scene Investigation Bingo

CODIS	class evidence	ballistics	handwriting	pH scale
antemortem	gas chromatograph	Edmond Locard	odontology	cold case(s)
latent	evidence		*mortis*	fingerprints
chromosome	height	barrier tape	impression evidence	caliber
forensic	coroner(s)	Latin	manner(s) of death	cast-off blood

Crime Scene Investigation Bingo

blood splatter	height	barrier tape	manner(s) of death	microscope(s)
ligature	medical examiner (ME)	chain of custody	gunshot residue	trajectory
caliber	suspect		latent	cast-off blood
luminol	cause of death	chromosome	CODIS	*rigor mortis*
cold case(s)	tool marks	Latin	coroner(s)	*modus operandi* (MO)

Crime Scene Investigation Bingo: Card No. 16

Crime Scene Investigation Bingo

caliber	postmortem	contamination	height	cast
CODIS	cold case(s)	impression evidence	trajectory	chain of custody
class evidence	*rigor mortis*		tool marks	barrier tape
DNA	forensic	FISH	ballistics	medical examiner (ME)
forgery	autopsy	handwriting	pH scale	suspect

Crime Scene Investigation Bingo: Card No. 17

Crime Scene Investigation Bingo

fingerprints	indented writing	cause of death	cold case(s)	ligature
modus operandi (MO)	caliber	forgery	latent	autopsy
class evidence	medical examiner (ME)		contamination	manner(s) of death
DNA	Edmond Locard	impression evidence	*rigor mortis*	gas chromatograph
tool marks	height	ballistics	postmortem	antemortem

Crime Scene Investigation Bingo: Card No. 18

© Barbara M. Peller

Crime Scene Investigation Bingo

latent	antemortem	height	barrier tape	indented writing
CODIS	pH scale	manner(s) of death	handwriting	trajectory
postmortem	cast		AFIS	fingerprints
gas chromatograph	tool marks	forgery	coroner(s)	contamination
mortis	microscope(s)	forensic	FISH	Latin

Crime Scene Investigation Bingo

indented writing	postmortem	pH scale	height	AFIS
blood splatter	cast-off blood	gunshot residue	forgery	ligature
suspect	chain of custody		luminol	Edmond Locard
DNA	FISH	*corpus delicti*	coroner(s)	tool marks
criminalistics	trace evidence	microscope(s)	*rigor mortis*	contamination

Crime Scene Investigation Bingo

CODIS	antemortem	gunshot residue	height	entomology (-gist)
suspect	contamination	cause of death	barrier tape	evidence
medical examiner (ME)	forensic		postmortem	ballistics
forgery	handwriting	tool marks	DNA	FISH
luminol	microscope(s)	Latin	caliber	coroner(s)

Crime Scene Investigation Bingo

mortis	gas chromatograph	contamination	odontology	caliber
ligature	pH scale	fingerprints	barrier tape	AFIS
cause of death	trajectory		evidence	chain of custody
tool marks	DNA	coroner(s)	Edmond Locard	cast
microscope(s)	autopsy	postmortem	medical examiner (ME)	gunshot residue

Crime Scene Investigation Bingo

blood splatter	postmortem	handwriting	odontology	Latin
antemortem	indented writing	forensic	CODIS	Edmond Locard
gas chromatograph	caliber		*corpus delicti*	evidence
medical examiner (ME)	microscope(s)	tool marks	autopsy	coroner(s)
entomology (-gist)	FISH	trace evidence	forgery	contamination

Crime Scene Investigation Bingo

blood splatter	indented writing	cast	postmortem	barrier tape
latent	Latin	gunshot residue	ligature	evidence
chain of custody	manner(s) of death		caliber	medical examiner (ME)
entomology (-gist)	*corpus delicti*	tool marks	autopsy	suspect
criminalistics	luminol	microscope(s)	pH scale	trace evidence

Crime Scene Investigation Bingo

luminol	gunshot residue	postmortem	ballistics	contamination
Edmond Locard	entomology (-gist)	CODIS	blood splatter	AFIS
suspect	barrier tape		*corpus delicti*	tool marks
manner(s) of death	DNA	trace evidence	microscope(s)	trajectory
Latin	cast	cause of death	cold case(s)	criminalistics

Crime Scene Investigation Bingo

contamination	postmortem	*corpus delicti*	ligature	manner(s) of death
forgery	pH scale	barrier tape	indented writing	blood splatter
entomology (-gist)	gas chromatograph		trajectory	luminol
caliber	odontology	DNA	microscope(s)	tool marks
chain of custody	cold case(s)	ballistics	trace evidence	criminalistics

Crime Scene Investigation Bingo

corpus delicti	cause of death	postmortem	indented writing	cast-off blood
entomology (-gist)	gas chromatograph	CODIS	tool marks	AFIS
impression evidence	trace evidence		microscope(s)	luminol
manner(s) of death	antemortem	gunshot residue	criminalistics	Edmond Locard
caliber	trajectory	contamination	*mortis*	chain of custody

Crime Scene Investigation Bingo

latent	indented writing	*rigor mortis*	postmortem	cause of death
cast-off blood	contamination	*corpus delicti*	forgery	trajectory
trace evidence	medical examiner (ME)		manner(s) of death	ligature
chain of custody	*mortis*	forensic	microscope(s)	tool marks
odontology	class evidence	caliber	criminalistics	entomology (-gist)

Crime Scene Investigation Bingo

contamination	indented writing	manner(s) of death	CODIS	class evidence
entomology (-gist)	forgery	gunshot residue	chain of custody	*mortis*
suspect	*corpus delicti*		AFIS	postmortem
cast-off blood	DNA	fingerprints	microscope(s)	tool marks
blood splatter	barrier tape	criminalistics	antemortem	trace evidence

Crime Scene Investigation Bingo: Card No. 29

Crime Scene Investigation Bingo

cast	postmortem	ligature	class evidence	suspect
Edmond Locard	manner(s) of death	gas chromatograph	trajectory	AFIS
criminalistics	autopsy		chain of custody	gunshot residue
entomology (-gist)	antemortem	indented writing	microscope(s)	*corpus delicti*
DNA	handwriting	trace evidence	contamination	fingerprints

Crime Scene Investigation Bingo: Card No. 30